GEOLOGIC GUIDE TO CLOUDLAND CANYON STATE PARK

By

Martha M. Griffin and Robert L. Atkins

Department of Natural Resources, Joe D. Tanner, Commissioner
Environmental Protection Division, J. Leonard Ledbetter, Director
Georgia Geologic Survey, William H. McLemore, State Geologist

Atlanta, Georgia
1983

GEOLOGIC GUIDE 7

Table of Contents

Era	Period	Epoch	Stages	Began years ago	Dominant life and important events
Cenozoic	Quaternary	Recent (Holocene)		11,000	man
		Pleistocene	Wisconsin	0.2 million	glacial
			Sangamonian	0.4 "	interglacial
			Illinoisian	0.6 "	glacial
			Yarmouthian	0.9 "	interglacial
			Kansan	1.4 "	glacial
			Aftonian	1.7 "	interglacial
			Nebraskan	3.0 "	glacial
	Tertiary	Pliocene		6 million	
		Miocene		22 "	
		Oligocene		37 "	mammals
		Eocene		54 "	
		Paleocene		62 "	
Mesozoic	Cretaceous			130 million	
	Jurassic			180 "	reptiles
	Triassic			230 "	
Paleozoic	Permian			280 million	
	Pennsylvanian			325 "	amphibians
	Mississippian			340 "	trees
	Devonian			400 "	grasses
	Silurian			450 "	fish
	Ordovician			500 "	
	Cambrian			580 "	invertebrates
Precambrian				1.8 billion	development of oxygen in atmosphere; appearance of eucaryotic cells
				2+ billion	life originated as a procaryote cell
				3+ billion	formation of ocean
				4+ billion	oldest rocks

GEOLOGIC TIME CHART

Note: This guidebook is designed to introduce the layman to basic principles of geology based on evidence of preserved and ongoing geologic processes in the exposed rocks of Cloudland

Canyon State Park. Because a detailed treatment of *stratigraphy* is beyond the scope of this guide, formation and member names have been deleted. The interested reader is directed to the bibliography for more complete information.

Words in italics throughout the text can be found in the Glossary.

INTRODUCTION

Scenic Cloudland Canyon State Park, located on Lookout Mountain in Dade County (Fig. 1), offers visitors one of the most spectacular panoramas in Georgia. Rugged and precipitous canyons, marked by dazzling waterfalls, plunge over 240 m (800 ft) to the floor of Lookout Valley. The park is located in the Cumberland Plateau that forms the extreme northwestern corner of the state. The rocks of this plateau represent the late Paleozoic Era (see Geologic Time Chart) and range from 340 to 280 million years old. These rocks, originally deposited in open sea and coastal environments, have been subsequently altered by deformation and erosion into the dramatic configurations that exist today.

The purpose of this guide is to interpret the exposed rocks of Cloudland Canyon as tangible and "readable" records of past events in the earth's history. Such an interpretation is largely based upon the concept that natural processes operating today are comparable to natural processes which operated in the past (*uniformitarianism*), or in simpler terms, "the present is the key to the past." Because this area represents deposits originally laid down adjacent to an ancient shoreline, examination of geologic processes at the modern coastline will serve as the key to identification of the rocks of the canyon.

In studying this area, relative dating is crucial because many physical events such as deposition of sediment, deformation, and canyon cutting can be identified. Proper chronological order can be determined through the concept that in a sequence of undeformed sedimentary rocks, the oldest beds are on the bottom and the youngest on top (*superposition*). At Cloudland Canyon, a walk down the trails to the canyon floor is an awesome descent through millions of years of geologic time, for each step downward is onto progressively older rocks.

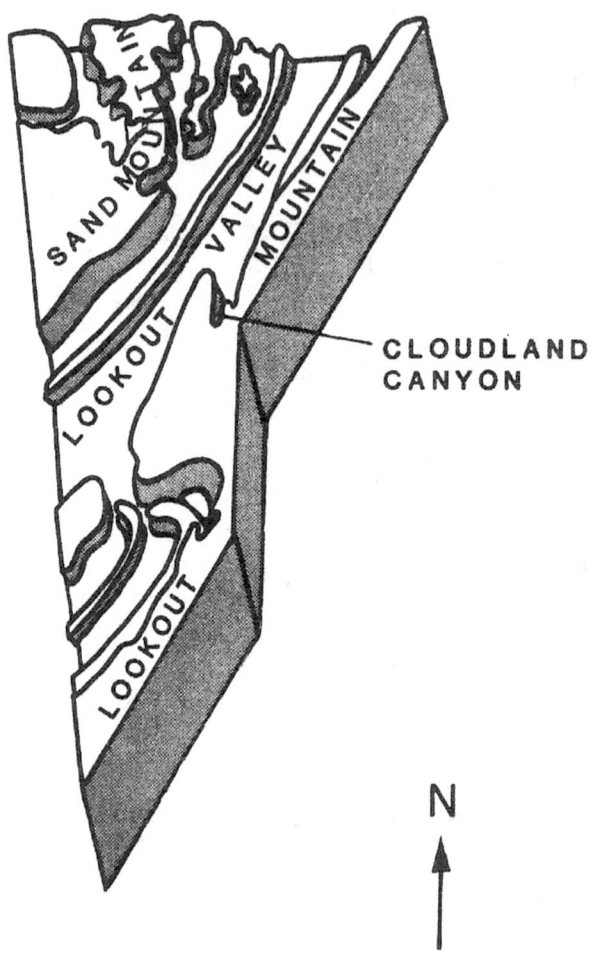

N

Observations using these basic geologic principles should increase your knowledge and understanding of the physical processes that, over the vast stretch of geologic time, have created the spectacular scenery of Cloudland Canyon. Increased knowledge of the geologic history of the canyon, as well as the geologic processes at work today, should enhance your pleasure during this visit to Cloudland Canyon as well as your commitment to the protection and conservation of Georgia's natural resources.

Late Paleozoic Geologic Setting

The rocks so dramatically exposed in Cloudland Canyon represent the Mississippian and Pennsylvanian Periods of the Paleozoic Era that are commonly grouped together as the *Carboniferous*, or coal-bearing sequence. During the Paleozoic Era, the terrain that now forms the southeastern United States lay in a vastly different setting. A worldwide warming trend, following a period of *glaciation*, melted most of the earth's extensive ice sheets and caused sea level to rise and flood the continent with shallow inland seas. West of the Appalachians, then a younger and far more rugged mountain chain, an elongate *basin* extended the length of the mountains. It was the gradual infilling of this trough, first with *limestone*, and, later with eroded *sediments* carried by streams from the uplands into the sea, that ultimately produced the Carboniferous rocks of the United States.

The warm, clear seas of this era supported a diverse population of sea life, as illustrated in Figure 2. From the calcium carbonate remains of these organisms was produced the limestone that represents an extensive Mississippian deposit in this area. Toward the latter part of the Mississippian Period, mud and sand were gradually deposited by rivers over the marine limestone (Fig. 3); this process continued through the Pennsylvanian Period. At the mouths of the rivers emptying into the basin, roughly triangular deposits of sediments, or *deltas*, formed from the river-transported sediments.

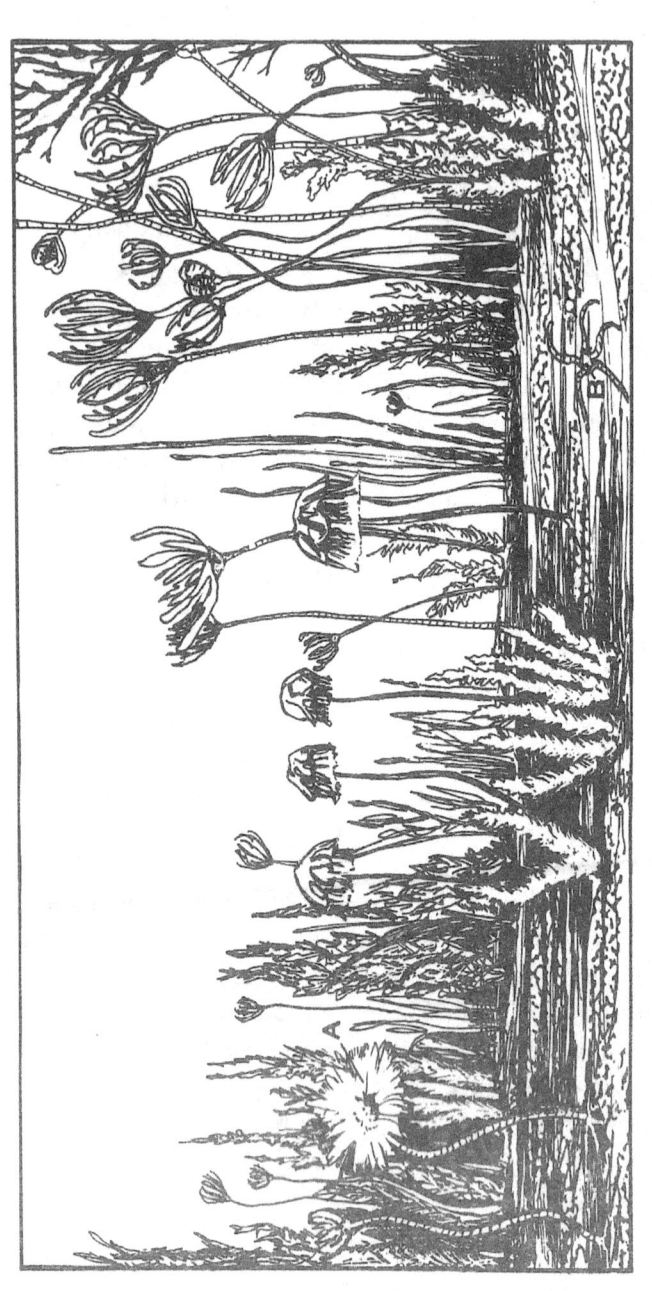

Figure 2. Life in a late Paleozoic sea. This Mississippian "sea lily" garden illustrates a crinoid assemblage on the sea bottom. Note blastoid (A) and brittle star (B).

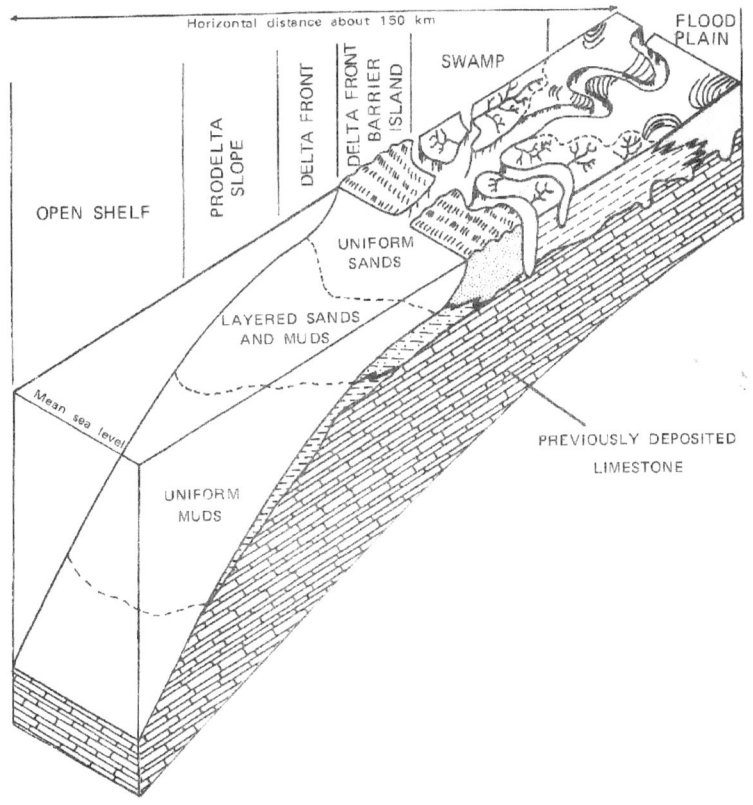

Figure 3. Delta deposition of mud and sand over previously deposited limestone. During delta growth, different environments migrated toward the sea.

Land plants had recently undergone an impressive explosion, during which the plants that provided most of the vegetation for coal deposits evolved (Fig. 4). The mild climate and low-lying terrain of the delta favored the development of extensive coal-forming swamps, vegetated by spore-bearing trees and ferns, and populated by amphibians, small reptiles, and insects.

Figure 5 illustrates the tracks of a small Pennsylvanian amphibian recently discovered very near Cloudland Canyon.

Thus the Carboniferous rocks of Georgia began as horizontal layers of sediments in or near a shallow Paleozoic sea, succeeding each other upward to form a thick sequence (Fig. 8A). A similar process is taking place today at the Mississippi Delta in the Gulf of Mexico. In the course of a very long period of time, the loose sediments were *consolidated* by pressure, resulting from the great weight of overlying sediments and accompanying process of cementation into the existing solid rocks, *shales, conglomerate, coal, sandstone,* and *limestone.* The history of these rocks had only just begun with the 60 million years it took for them to form.

Figure 4. Coal-forming Carboniferous land plants.

GEOLOGIC TRAIL

The geologic trail (Fig. 6) is designed to acquaint the curious visitor with visible evidence of both the geologic history and present day geologic processes of Cloudland Canyon. The trail, which begins at The Point, consists of eleven stations and covers close to 10 km (6 mi). Visitors moving at a leisurely pace may require up to 6 hours to complete the hike and will probably wish to take along food and drink. (If you prefer a short hike of 1 to 2 hours, end your trip at the first waterfall and just read the rest of the guide).

VISITORS MUST REMAIN ON THE TRAILS AT ALL TIMES AND KEEP CHILDREN UNDER CLOSE SUPERVISION TO INSURE A SAFE TRIP.

Figure 5. Trackway of a small Pennsylvanian
amphibian, _Cincosaurus cobbi_, discovered
on Lookout Mountain. (Photo courtesy of
Bill Schneck.)

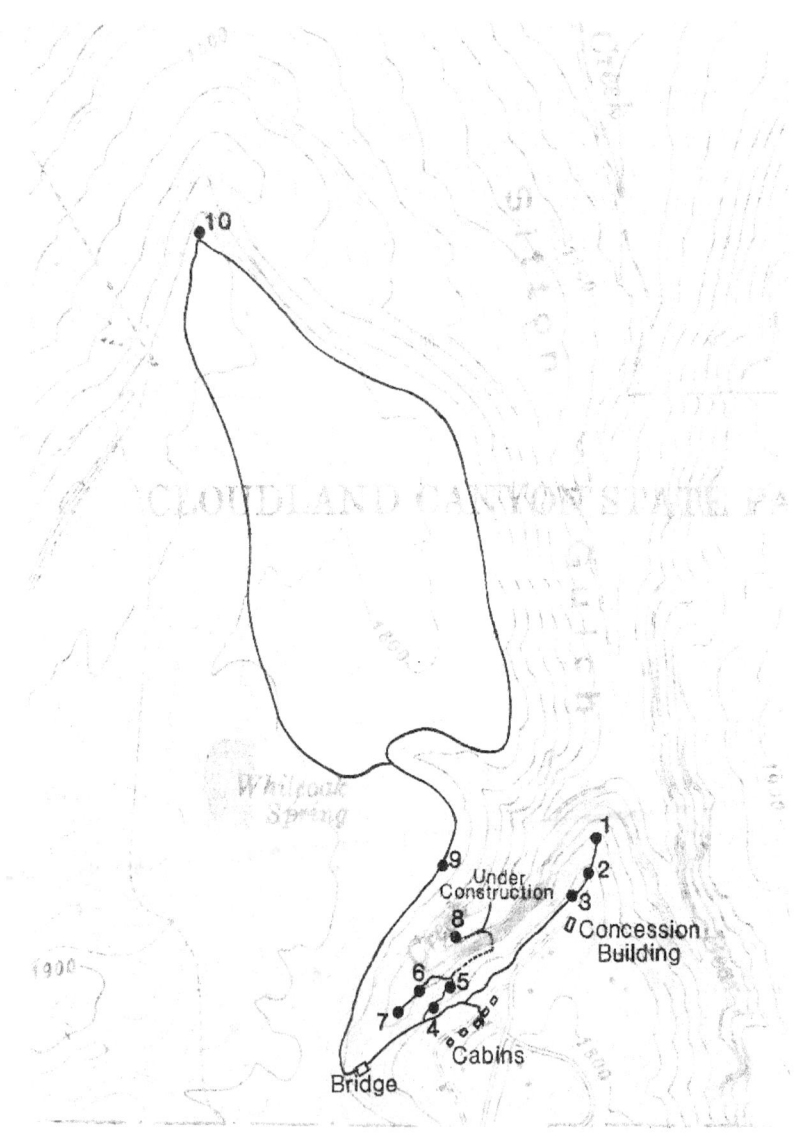

Figure 6. Geologic Trail Map.

STATION #1: The Point, East Rim Trail

From your vantage point atop the sandstone cliffs of Lookout Mountain, you stand some 240m (800 ft) above the limestone floor of Lookout Valley (Fig. 7). The deep gorge known as Cloudland Canyon has been carved by Daniel Creek to the west and Bear Creek to the east. How were such great thicknesses of rock removed?

During late Paleozoic time, these rocks were deformed by enormous lateral pressure from the east into *folds*, or wave-like contortions in the rocks. The Carboniferous rocks were folded into *anticlines*, or upwarps, and *synclines*, or downwarps (Fig. 8). These large fold structures are too extensive to be seen in their entirety from any one point in the park.

Figure 7. Lookout Mountain and Lookout Valley from the Point. Note that the sandstone of Lookout Mountain forms steep bluffs, whereas the limestone forms the gentle slopes which open into Lookout Valley.

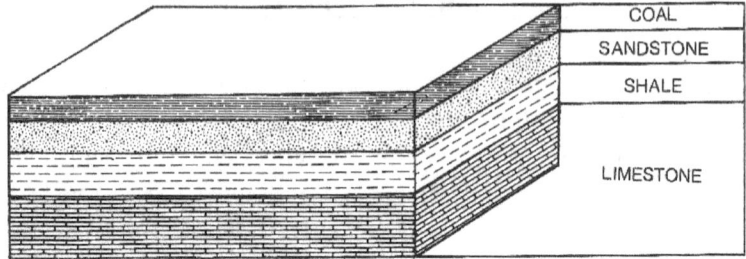

COAL
SANDSTONE
SHALE
LIMESTONE

A.   Depositional sequence.

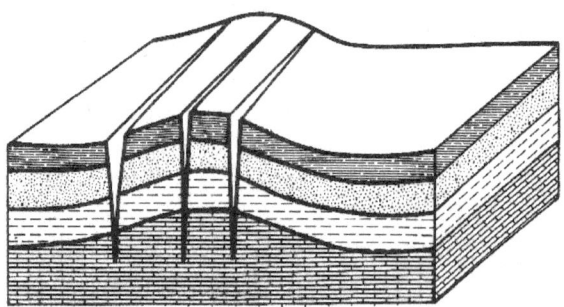

B. Deformation into anticlines and synclines. Note greater number of fractures in anticline.

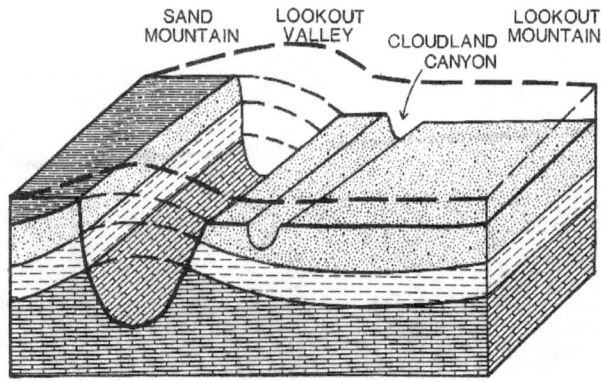

SAND
MOUNTAIN

LOOKOUT
VALLEY

CLOUDLAND
CANYON

LOOKOUT
MOUNTAIN

C. Erosion of structure resulting in valley and ridge.

Figure 8.   Generalized cross-section of the Cloudland
Canyon area. Note diagrammatic represen-
tation of the (A) depositional sequence,
(B) deformation, and (C) erosion of the
Cloudland Canyon area.

In the course of the compressive event that folded the Paleozoic rocks into anticlines and synclines, the rocks were also broken by systems of vertical fractures. If rock slippage occurs along these fractures, they are known as *faults*. If no appreciable movement has taken place along these fractures, they are known as *joints*. The long smooth surfaces on the canyon walls are bounded by joints. You can see that the sandstone has been broken by many *joint sets*, or systems of parallel joints (Fig. 9). Because running water invariably seeks the course of the least resistance, these joint sets have determined the stream patterns that eventually cut through the resistant sandstones and conglomerates to form the spectacular canyons of the park.

During the folding event that deformed these rocks into broad folds, more stress fractures developed in the anticlines than in the synclines (Fig. 8B). With successive cycles of erosion these fractures, or joints, have caused the anticlines to erode at a faster rate than the synclines. Consequently, the fractured anticlines eventually eroded completely to form valleys, whereas the synclines resisted erosion to stand above the valleys as steep sided mountains (Fig. 8C).

Forming the bluff of Lookout Mountain is a hard, durable sandstone. The nature of the *bedding*, or division of sedimentary layers, suggests deposition in a delta environment, as do the great thickness of the sand and the presence of coal. Careful inspection of the rocks at your feet will reveal well preserved ripple marks (Fig. 10) similar to those forming at the present day coast.

Figure 9. West Canyon wall. Note joint patterns of
vertical fractures and the variability of
sandstones exposed on the wall. The shale
layers are marked by pine tree growth.

Figure 10. Preserved ripple marks in sandstone, East
Rim Trail. These ripples were formed by
moving water at shallow depth.

Notice that there are both steep and gentle slopes on the walls of the two canyons that open into the broad valley below you. Do you suppose that the sandstone, shale, and limestone which formed at the Paleozoic shoreline have weathered uniformly? As you probably guessed, these rocks have eroded at widely varying rates depending on their hardness and resistance to weathering; this phenomenon is known as *differential erosion*. The broad valley has been carved out of limestone that is water soluable and, therefore, easily weathered. The rugged canyons of the park were carved from the more resistant sandstone, conglomerates, and shales, along vertical fractures which allowed water to migrate through the rock.

STATION #2: Overlook at Picnic Shelter, East Rim Trail

Look across the canyon and note the variability of the exposed sandstones of the canyon wall (Fig. 9). The upper, massive, bluff-forming unit grades downward into a less resistant unit of sandstone, marked by occasional layers of shale. The sandstone layers are sheer and free of vegetation, whereas the shale layers are characterized by the growth of pine trees in their more gentle slopes and more nourishing soil.

A variety of sandstones may be seen in the rooks cemented into the guard rail supports. You should see many rocks containing pebbles as well as sand; these are known as conglomerates, since they contain a "conglomeration" of sediment sizes. Look for an example of *graded bedding*, as illustrated in Figure 11; if you shake up a bottle of water, pebbles, and coarse-to-fine sand, the sediments will settle out in just this pattern, that is, coarse below and fine above. Graded bedding can occur in thicknesses ranging from a few centimeters to hundreds of meters, depending on the water depth and the quantity of waterborne sediment.

STATION #3: <u>Overlook at Telescope, East Rim Trail</u>

Close inspection of the west canyon wall will reveal several bowl-shaped depressions at different levels (Fig. 12). These bowls are easily recognized when marked by "wet weather" waterfalls, fed by ground water trickling down through joints in the rocks. Over time, the circular motion of sediment-laden water at the base of the falls has carved these bowls into the sandstone and shale. Again and again, the rock overhang produced by the undercutting of turbulent water has collapsed, thus deepening and enlarging these features.

Figure 11. Small-scale graded bedding. This rock, part of the guard rail near The Point, illustrates layering based on grain size.

Notice that the lower walls of the canyon are littered with broken rocks; such an accumulation of rock fragments at the base of a cliff is known as talus. (This talus slope is heavily vegetated and will be more difficult to identify during warm seasons.)

Near the top of the opposite canyon wall there is a group of huge, roughly cubic boulders (Fig. 13) that appear to be about to slide over the precipice. Remember these boulders well, for you will examine them at close range when you reach the opposite side of the canyon.

CONTINUE ON THE WATERFALL TRAIL, FOLLOWING THE SIGNS DOWN THE PATH TO THE FALLS.

Figure 12. Bowl-shaped depressions on west canyon
wall. Wet weather waterfalls have carved
these concave depressions through the
cutting action of turbulent, sediment-
laden water. Note that the darker shale
layer has been incised more deeply than
the sandstone.

Figure 13. Sandstone boulders, West Rim Trail. These
massively bedded sandstone boulders appear
to be a continuation of the filled stream
channel deposit on the east canyon wall.

STATION #4: <u>Shallow Water Crossbeds, Waterfall Trail</u>

At this stop, just beyond the wooden steps, you can again examine at close range the sedimentary structures that indicate a coastal environment. These thin sandstone layers, lying at a gentle slope, indicate deposition by wave or tidal action (Fig. 14); such sedimentary structures are known as *crossbeds*. During water or wind transport, successive layers of sand grains are carried up the slope, then down the steep side of the sand layer. A storm-cut dune on Sapelo Island, Georgia reveals a similar layering, or crossbedding, of the sands (Fig. 15).

As you continue down the trail to the next stop, pause and look back and upward. You should be able to recognize the *contact*, or change in bedding, from the uppermost, massively bedded sandstone to the underlying cross-bedded sandstone unit.

Figure 14.   Low angle crossbedding, east canyon wall. Note preserved layering of water-deposited sand.

As you continue down the trail to the next stop, pause and look back and upward. You should be able to recognize the *contact*, or change in bedding, from the uppermost, massively bedded sandstone to the underlying cross-bedded sandstone unit.

Figure 15. Crossbedding in modern day beach sands. This storm-cut dune on Sapelo Island, Georgia, reveals crossbedding similar to that preserved at Cloudland Canyon.

STATION #5: <u>Rock Overhang, Waterfall Trail</u>

Does the shape of this huge rock overhang and the strange markings on its underside provide a clue to its origin? The bluff-forming sandstone of the park is characterized by this massive, sand-filled ancient stream channel (Fig. 16). Although the *channel fill* has been breached by a combination of fracturing and erosion, its original course appears to have continued across the canyon; this channel can be inferred by aligning exposures of this sandstone from either side of the canyon. Close examination of the base of the fill reveals a deposit of coarse materials, or *channel lag*, that accumulated on the stream bed. The lack of layering in the sandstone indicates that this ancient channel was rapidly filled with sand after the stream changed course.

Figure 16. Sand filled stream channel deposit. The channel fill visible to the left is the result of rapid infilling with sand after the stream changed course. The thin layers, or crossbeds, of the lower unit are a result of wave or tidal action.

The bottom of the rock overhang is marked by small, irregular structures known as *sole marks*. These result when currents scour out shallow depressions in the channel bottom. When the current slows, the depressions are filled with sediment. Preserved sole marks are important to geologists as indicators of stream flow direction.

STATION #6: Shale Outcrop, Waterfall Trail

As you approach the base of the canyon, you should notice an abrupt change in the rocks exposed on the canyon wall (Fig. 17). The bedding is considerably thinner and dark brown to black in color, and the trail is littered with broken, dark chips. Pick up one of the chips and note that this shale, or clay hardened into rock, is much softer and easier to erode than the sandstone. For this reason, the more resistant sandstone has formed an overhang above the shale. The silts and clays that were consolidated into this shale were likely deposited at the seaward portion of a delta. Close inspection of the shale reveals many thin layers clay, which settled out in quiet water between high and low tides, or at "slack tide." This process of clay accumulation is taking place today in the broad marshes, lagoons, and tidal flats that lie between the mainland and the barrier islands of Georgia.

STATION #7: Base of Upper Daniel Creek Waterfall

Here, in the lower reaches of the canyon, are found the alternating layers of sandstone and shale so well exposed at this 15.2m (50 ft) waterfall. Note the small, filled stream channel in the shale layer to the left of the falls, the small talus slope to the right, and the coarse sand which has accumulated at the water's edge. The filled stream channel is a record of past erosion, while the talus slope is a reminder of present day erosion. The coarse grains rimming the splash pool have been weathered out of the ancient sandstone rocks for yet another journey to the sea.

Figure 17. Shale outcrop, lower east canyon wall. These thin, even layers of consolidated silt and clay suggest that the sediments settled out during consecutive "slack tides."

Cloudland Canyon is a dramatic example of the power of a stream to cut downward through rocks broken by fractures. One must realize that a stream is not just flowing water, but a system of moving water and sediment combined. The abrasive action on a stream channel as gravel and sand are swept along the bottom is not unlike that of a sand blaster.

Figure 18. Upper Falls of Daniel Creek. The circular motion of water and sand at the base of the falls has carved a depression in the lower shale, producing a sandstone overhang.

When a stream cuts down to a hard, resistant rock layer, such as sandstone, the slope of the stream is interrupted and waterfalls result. Here at the Daniel Creek Waterfall, for example, a resistant sandstone forms an overhang over the weaker shale (Fig. 18). As illustrated by the many waterfalls in the park, a stream may flow over more than one resistant rock layer.

Where the hydraulic activity of the stream is strong, such as at the base of a fall, its cutting action is particularly effective.

TO PROCEED TO STATION #8, RETURN UP THE WOODEN STAIRS, THEN TAKE THE TRAIL DOWN TO THE LOWER FALLS.

STATION #8: <u>Base of Lower Daniel Creek Waterfall</u>

At the lower falls of Daniel Creek, we see an even more spectcular 27m (90 ft) waterfall, cascading over a hard sandstone outcrop. Again, even layers of slack-tide shale deposits are visible under the sandstone overhang, broken rocks circle the splash pool, and large crossbeds are visible on the canyon walls.

The focal point of the scene, however, is the massive sandstone boulder at the base of the falls (Fig. 19). Based on bedding and grain size evidence this "big-as-a-house" rock can only be a huge, broken section of the channel fill previously examined.

Now that you are well into the depths of the gorge, pause once again to consider the layering of these ancient rocks. You may recall that the Pennsylvanian delta muds and sands (325-280 million years old) overlie the Mississippian open sea limestone (340-325 million years old). The relative ages of the rocks have been established through identification of marine fossils in the limestone and plant fossils in the shale.

Compare the stratigraphy of this ancient, preserved Paleozoic shoreline with that of the modern day Georgia coast, as illustrated in Figure 20. Note that at Cumberland Island thick deposits of barrier island sands overlie older marsh-lagoon clays. This sequence of unconsolidated sediments rests upon older river deposits, which in turn overlie ancient limestone, formed about 50 million years ago when sea level was higher. In spite of obvious differences resulting from a shoreline associated with a shallow inland sea as compared to a shoreline at the open ocean, the same near shore processes which resulted in the deposition of the Carboniferous rocks are at work on the present day coast. Thus it is said that "the present is the key to the past."

RETURN TO THE BRIDGE TRAIL TO CONTINUE HIKE; ALLOW ABOUT THREE HOURS TO COMPLETE THE WALK.

Figure 19. Lower Falls of Daniel Creek. The huge
boulder is a broken remnant of the channel
fill examined at Station #5. Note figure
at left for scale.

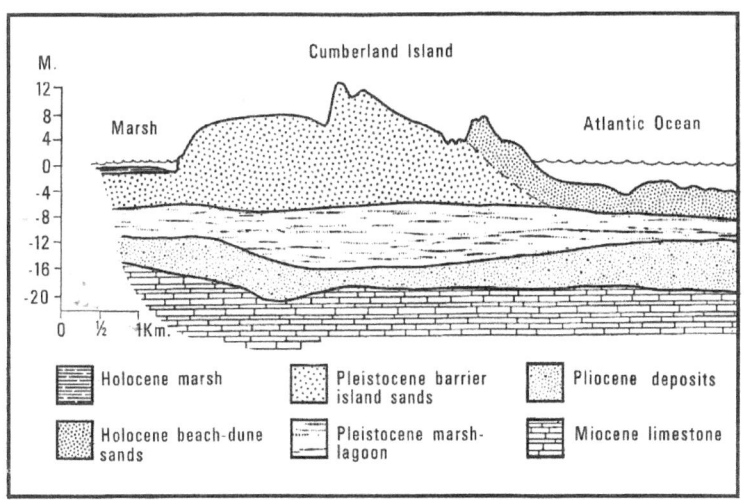

Figure 20. Sediment layers of Cumberland Island, Georgia. Due to rising sea level, the barrier island sands have moved landward and buried the old marsh clays and the older river deposits. The entire sequence rests on the ancient limestone of Georgia's major aquifer.

STATION #9: <u>Boulders on West Rim Trail</u>

These huge sandstone boulders were consolidated into hard rock while buried and are now adjusting to a different environment at the earth's surface. This adjustment, known as *weathering*, is the breaking down of rock by both physical and chemical means.

Chemical weathering, in which disintegration results from a chemical reaction between the elements in the atmosphere and those in the rocks, is caused by water and by organic acids produced by plants and animals. Water is the principal agent of chemical weathering in the warm and humid climate of Georgia. Given enough time, water dissolves minerals and erodes them away.

Mechanical weathering is the breakdown of rocks into smaller fragments by physical means, for example, the boulders that have fallen from the rim. The most important types of physical weathering are (1) ice wedging, in which freezing water expands in cracks and wedges the rock apart, and (2) sheeting, the process by which once buried rock expands in response to removal of overlying pressure, causing horizontal fracturing. Physical weathering may also be caused by pressure from growing roots and burrowing animals, as well as the activities of man.

Physical weathering aids chemical weathering by breaking up the rocks and exposing more surface area to water and air. A good example of physical weathering can be seen where trees have taken root in cracks in the rocks (Fig. 21). As the roots grow, they exert great pressure on the rock, pushing it apart to expose more surface area to chemical weathering. What does the size of the trees growing in the joints indicate about the speed with which these boulders are moving toward their inevitable fall into the canyon?

Joints greatly influence the weathering of rock bodies by (1) allowing the breakage of large blocks of rocks into smaller ones, thereby greatly increasing the surface area available for chemical reactions, and (2) acting as a channel way through which water can penetrate.

A look back to the east canyon wall should be your clue to the origin of these impressive boulders, for here lie the broken remains of the massive channel fill examined at Stations #5 and #8.

STATION #10: <u>Lookout Valley Overlook, West Rim Trail</u>

The geologic trail walk has dealt primarily with the delta sandstones and shales that formed in or near the warm, shallow waters of a Paleozoic sea, for these are the rocks so well exposed on the canyon walls. But what of the extensive ancient limestones that underlie these shales and sandstones? And what of the coal-bearing, delta marsh deposits that buried the barrier island and delta sands?

In Cloudland Canyon, the ancient Mississippian limestone is heavily vegetated where it gently slopes from the canyon walls and opens into the broad expanse of Lookout Valley. Remember that this *fossiliferous limestone* formed out of the remains of the rich and varied sea life of the Paleozoic sea that once flooded our continent.

Many marine plants and invertebrate animals extract calcium carbonate from sea water and use it to construct their shells and hard parts. When the organisms die, their shells accumulate on the seafloor and, over a long period of time, build up a deposit of limestone consisting of shells and shell fragments. Limestone produced in this way can form thick and widespread deposits extending over thousands of square kilometers. The ancient limestone *aquifer* that serves as the principle source of ground water for the Georgia Coastal Plain is a similar deposit (Fig. 22). Although the ocean waters off the Georgia coast are now too cool and turbid to support major limestone forming life forms, fossiliferous limestone is today forming in the warm, clear waters of the Florida Keys.

The coal bearing sandstone of this region cannot be seen in the park, for erosion has removed the delta plain deposits which originally overlaid the present sandstone bluff (Fig. 8C). On clear days, the scars of coal mines may be seen on Sand Mountain, a testimony to the extensive mining industry that once flourished in this region (Fig. 23).

Figure 21. Physical weathering; West Rim Trail.
These boulders, produced by fracturing of
the channel fill, are undergoing physical
weathering through tree growth along the
joints.

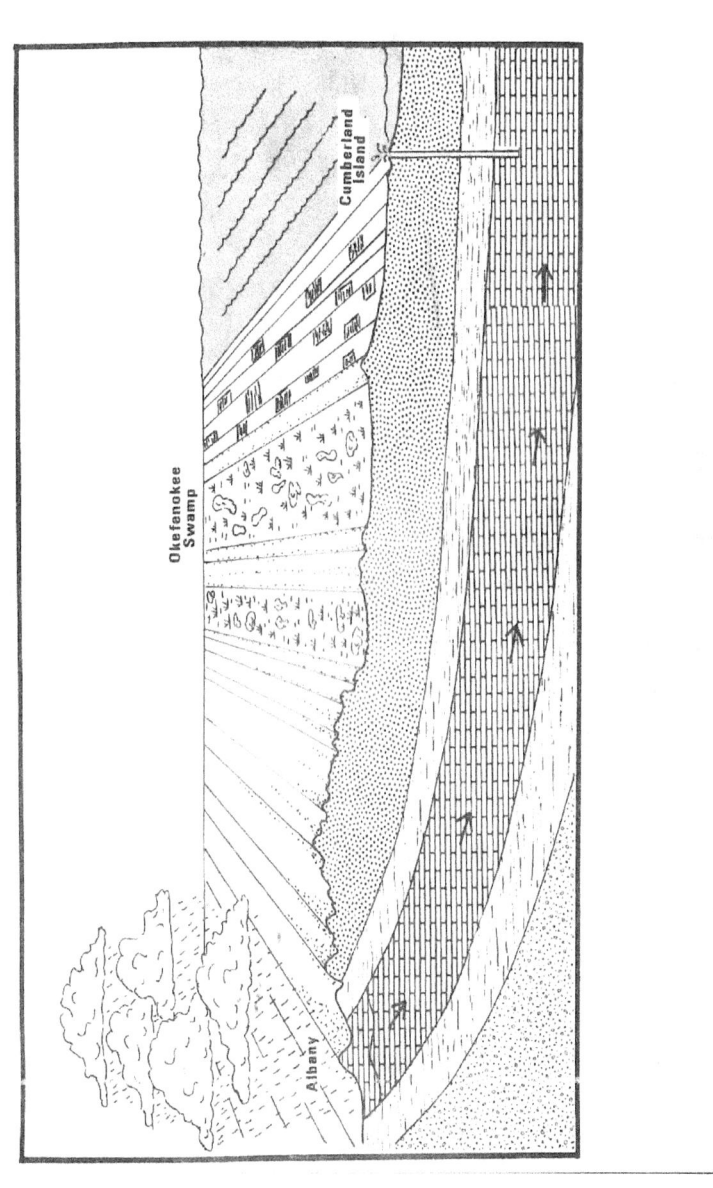

Figure 22. Schematic diagram illustrating the principal aquifer of the Georgia Coastal Plain. Arrows indicate movement of ground water.

Figure 23. Coal mining activity in Dade County, Georgia, circa 1903. Pictured are the coke ovens of the Georgia Iron and Coal Co. in Cole City. Cole City is an abandoned coal mining town due west of New Hope, Georgia, in Nickajack Cove.

Geologists believe that the nearby coal bearing sandstones resulted from peat beds formed in delta plain marshes. Coal originates from plant material that flourished in ancient swamps typically found on low lying coastal plains. A modern example is the present Okefenokee Swamp of Georgia, where lush vegetation has produced thick layers of peat (partially decomposed plant material). In such an environment, the layer of peat may be covered with sand and mud in the event of a sea level rise (Fig. 24). Under the resulting pressure from overlying sediments, peat eventually is compressed and transformed into coal.

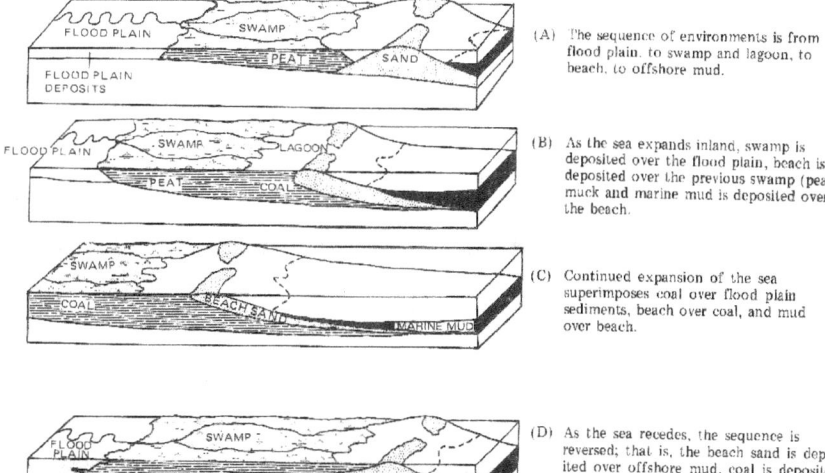

(A) The sequence of environments is from flood plain, to swamp and lagoon, to beach, to offshore mud.

(B) As the sea expands inland, swamp is deposited over the flood plain, beach is deposited over the previous swamp (peat) muck and marine mud is deposited over the beach.

(C) Continued expansion of the sea superimposes coal over flood plain sediments, beach over coal, and mud over beach.

(D) As the sea recedes, the sequence is reversed; that is, the beach sand is deposited over offshore mud, coal is deposited over beach sand, and flood-plain sediments are deposited over coal.

Figure 24.   Coal formation resulting from swamp burial. This schematic diagram illustrates how a swamp can be buried to form a coal deposit.

CONCLUSION

Within the confines of Cloudland Canyon State Park, you have today witnessed the cumulative effect of hundreds of millions of years of geologic activity. You have examined traces of the genesis of these rocks preserved in structures such as ripple marks and bedding. You have viewed the extensive fractures produced in these ancient rocks from enormous pressure deep in the earth. You have watched the dramatic effects of the downcutting action of streams, as well as the subtle but relentless advance of physical and chemical weathering. While geologic forces will forever alter the face of the earth, a greater understanding and appreciation of the physical processes responsible for this canyon should increase both your pleasure in its beauty and your commitment to its conservation.

ACKNOWLEDGEMENTS

The authors wish to express their thanks to Bill Thoman, Region 1 Naturalist, Georgia Parks and Historic Sites; Mildred W. Graham, Science Education Dept., Georgia State University; and Jim Sadd, Geology Dept., Georgia State University, for encouragement, advice, and critical review.

Publications Editor/Coordinator: Eleanore Morrow

BIBLIOGRAPHY

Butts, Charles, and Gildersleeve, Benjamin, 1948, Geology and mineral resources of the Paleozoic area in northwest Georgia: Ga. Department of Mines, Mining and Geology, Bulletin 54, Atlanta.

Chowns, T.M., 1978, Molasse sedimentation in the Silurian rocks of northwest Georgia, in Sedimentary environments in the Paleozoic rocks of northwest Georgia, Guidebook 11, Geological Survey of Georgia, Atlanta.

Croft, M.G., 1964, Geology and groundwater resources of Dade County, Georgia: Georgia Geologic Survey, Information Circular 26, Atlanta.

DeSitter, L.U., 1964, Structural Geology: McGraw-Hill, New York.

Griffin, Martha M., 1982, Geologic guide to Cumberland Island National Seashore: Georgia Geologic Survey, Atlanta, Geologic Guide 6.

Levin, H.L., 1978, The earth through time: Saunders Company, Philadelphia.

McCallie, S.W., 1904, Coal deposits of Georgia: Geological Survey of Georgia, Atlanta, Bulletin 12.

Schneck, William M., 1982, The first record of Carboniferous amphibian trackways in Georgia: unpublished paper submitted to the Phi Beta Kappa Faculty Group, Georgia State University Atlanta.

Thomas, William A., and Cramer, Howard R., 1979, Mississippian and Pennsylvanian systems in the United States:

U.S. Geological Survey Professional Paper 1110-H, Washington, D.C.

GLOSSARY

ANTICLINE: A fold in which the limbs dip away from the axis (upwarp). When eroded, the oldest rocks are exposed in the central core.

AQUIFER: A permeable zone below the earth's surface through which ground water moves.

BASIN: A depressed area that serves as a catchment for sediments.

BEDDING: Existence of planes of separation dividing sedimentary rocks.

CARBONIFEROUS: Period equivalent to combined Mississippian and Pennsylvanian. (See geologic time chart.)

CHANNEL FILL: Deposits representing sedimentation in an abandoned stream channel.

CHANNEL LAG DEPOSITS: Relatively coarse materials that have been sorted out in the normal processes of a stream and left as a residual accumulation on the bottom.

CHEMICAL WEATHERING: The breakdown of rock resulting from chemical reactions of elements in the atmosphere with those in the rocks.

COAL: A naturally occurring, rocklike derivative of peat, compressed and altered to material with increasing carbon content.

CONGLOMERATE: A sedimentary rock containing rounded fragments of pebbles, cobbles, or boulders.

CONSOLIDATION: Any or all of the processes by which loose, soft, or liquid earth materials become firm and coherent.

CONTACT: The surface separating two different rock bodies.

CROSSBEDDING: Stratification inclined to the original horizontal surface upon which the sediments accumulated.

CUMBERLAND PLATEAU: The eastern escarpment of the Appalachian Plateau, delineated by Lookout and Sand Mountains in Georgia and Alabama.

DELTA: A deposit of sediment, roughly triangular, at the mouth of a river as it empties into the sea.

DELTA PLAIN: Plains formed by the accumulation of silt at the mouths of streams, or by overflow along their lower courses.

DIFFERENTIAL EROSION: Variations in rates of erosion on different rock-masses. As a result, resistant rocks form steep slopes and non-resistant rocks form gentle slopes.

EROSION: The process that loosens and moves sediment to another place on the earth's surface.

FAULT: A fracture or fracture zone along which there has been displacement of the sides relative to one another.

FOLD: A bend or flexure in a rock.

FOSSILIFEROUS LIMESTONE: Limestone containing organic remains, such as shells.

GLACIATION: A part of geologic time when a large portion of the earth was covered by ice sheets.

GRADED BEDDING: A type of bedding in which there is a characteristic decrease in grain size from bottom to top.

JOINT: A fracture in a rock along which there has been no appreciable displacement.

JOINT SET: A system of parallel joints.

LIMESTONE: A sedimentary rock composed primarily of calcium carbonate.

MECHANICAL WEATHERING: The fragmentation of rocks by physical forces.

PEAT: Partially decomposed plant material, considered an early stage in the development of coal.

SANDSTONE: A sedimentary rock composed mostly of sand sized particles cemented by calcite, silica, or iron oxide.

SEDIMENT: Material, such as gravel, sand, mud, or lime, that has been transported and deposited by wind, water, ice or gravity.

SEDIMENTARY ROCK: Rock formed from the accumulation and consolidation of sediment.

SHALE: A fine-grained, clastic (fragmented) sedimentary rock formed by the consolidation of clay and mud.

SOLE MARKS: Marks produced as a result of the current flowing over and scouring the sediment surface.

STRATIGRAPHY: Formation, composition, sequence and correlation of layered sedimentary rocks.

SUPERPOSITION: The principle which states that in a sequence of undeformed sedimentary rocks, the oldest beds are on the bottom and the youngest are on the top.

SYNCLINE: A fold (downwarp) in which the limbs dip toward the axis and the youngest beds are in the central core after erosion.
Rock fragments that accumulate in a pile at the base of a ridge or cliff.

UNIFORMITARIANISM: The theory that the earth's surface and geology are a result of natural processes, many of which are operating at the present time.

WEATHERING: The chemical and mechanical breakdown of rock with little or no transportation of the loosened or altered material.

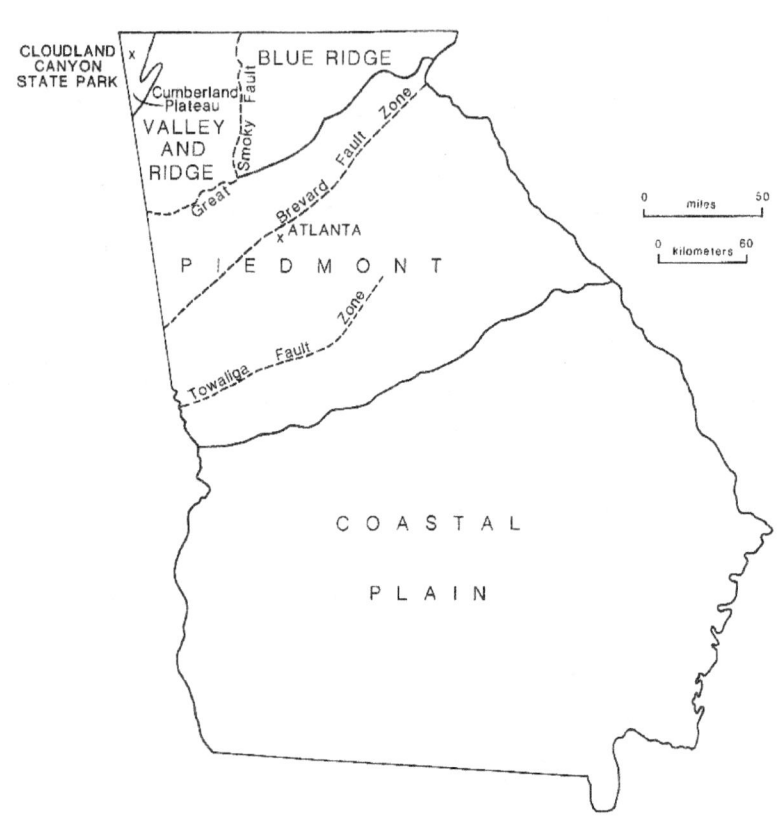

CLOUDLAND
CANYON
STATE PARK

BLUE RIDGE

Cumberland
Plateau

VALLEY
AND
RIDGE

Great Smoky Fault

Brevard Fault Zone

x ATLANTA

P I E D M O N T

Towaliga Fault Zone

0   miles   50

0   kilometers   60

C O A S T A L

P L A I N

GENERALIZED PHYSIOGRAPHIC PROVINCES OF GEORGIA

NOTES